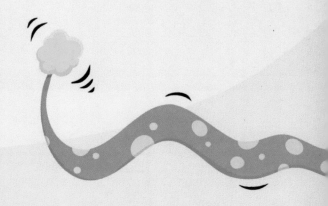

For Vega and Jedson,
You are and always will be my inspiration.
- Dad

For Oscar,
May your life be filled with colourful adventures, always.
- Nemy

James Jefferson
Illustrations by Nemy Arthur

I dress up in my favourite spacesuit whenever it's time to watch a space launch with my family.

Every night as I get ready for bed, I look out the window at the Moon and all the stars. I often wonder what it would be like to go to space.

Each night when the skies are clear, I gaze into the telescope and wonder what adventures await me.

The Moon has no gravity, so when I was on the surface, I would jump up and down. I'd bounce over its craters in one giant leap!

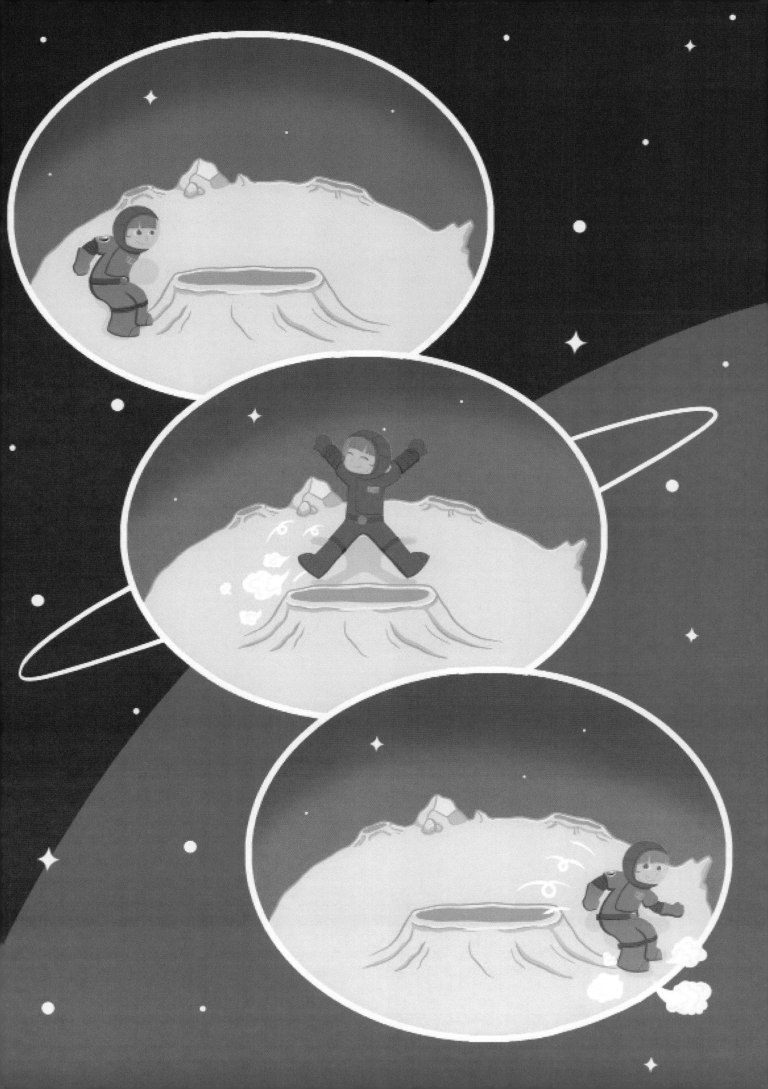

Mars has the largest volcano in our solar system. It is called Olympus Mons. Looking up from the bottom of the volcano, I would start to climb one small step at a time.

While exploring Jupiter's moon Europa, I would run in and out of water geysers. I'd make sure to avoid getting my spacesuit wet.

In my rocket ship, I would surf the beautiful rings of Saturn, dodging the planet's huge icy rocks.

If I want to go to space, I need to work hard at school, listen to my teachers and do my homework.

I should try to eat five fruits and vegetables each day and drink plenty of water. This will help me concentrate and think clearly.

I continue to exercise.

...and get plenty of rest.

Did you know?

The closest galaxy to us is the Andromeda Galaxy — it's estimated at 2.5 million light-years away.

On 16th June, 1963, Soviet Cosmonaut Valentina Tereshkova became the first woman to travel into space. After 71 hours, Tereshkova piloted her spacecraft back into the Earth's atmosphere and then parachuted to Earth.

elliptical

spiral

irregular

There are three main types of galaxies: elliptical, spiral & irregular

On July 20, 1969, Neil Armstrong became
the first human to step on the moon.
He was later joined by Edwin (Buzz)
Aldrin. They stayed on the Moon
for 21hrs 36mins.

There is an asteroid belt between Mars and
Jupiter. This is where most of the asteroids
in our solar system are found.
Also located in this region is
the dwarf planet Ceres.

Pluto

Pluto was once the 9th planet
in our system. However, in 2006 it was relegated
to a dwarf planet as it didn't meet the new criteria
set out the IAU (International Astronomical Union).
Pluto didn't have the gravitational pull needed to clear
the area around its orbit of the Sun.

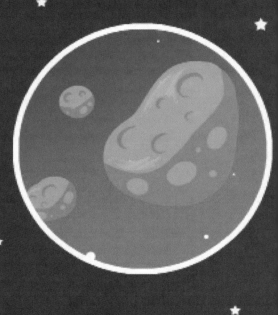

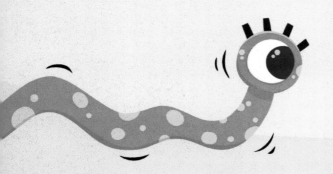

Printed in Great Britain
by Amazon

40808100R00030